Amara's Adventures

AMARA AND THE SECRET WORLD OF BUTTERFLIES

By Nola D Oracle

Illustration by Casielle Santos-Gaerlan

This special book is dedicated to all the kids who enjoy going on exciting journeys, learning, and discovering along the way! A big shout-out to the Florida Museum of Natural History's McGuire Center for Lepidoptera and Biodiversity in Gainesville, Florida!

A super special thank you to the fantastic Jaret C. Daniels! You are truly appreciated for spending so much time researching, curating, and sharing the wonderful world of butterflies for young minds to explore and enjoy.
Great job!

Peace and Love, Friends!
Today, we're going on an
exciting adventure to learn
all about the amazing secrets
of butterflies. I'll be your guide,
so come along; let's explore
and discover together!

Butterflies have a very special life cycle! Each stage is full of surprises. Let's go explore and see how they grow and change!

A butterfly's life cycle begins with tiny eggs. They are so small you can barely see them, and they usually hatch in just 3 to 7 days.

The second stage of a butterfly's life cycle is the caterpillar, also called the larva. The caterpillar spends 2 to 3 weeks eating and growing as fast as it can!

The third stage is called the pupa, or chrysalis. Inside the chrysalis, the caterpillar changes into a butterfly. This magical stage usually takes about 1 to 2 weeks.

The last stage is when the butterfly becomes an adult. It spreads its new wings and flies into the world to explore new places!

Butterflies come in all colors and shapes. Some have fun patterns.

Butterflies drink sweet nectar from flowers and also enjoy the juices from fruits.

Butterflies spread pollen as they fly from flower to flower, helping plants grow big and healthy.

Butterflies help flowers
make food, which gives us
fruits and veggies to enjoy!

Lots of special plants love to attract butterflies, and we can plant them to invite butterflies to visit our gardens! Let's go discover them!

Butterflies love milkweed plants! They're a favorite playground for colorful monarchs.

Lantana plants are a
butterfly magnet with their
sweet nectar.

In the fall, aster plants shine bright and attract butterflies when other flowers are slowing down.

With their sweet smell and
cone-shaped flowers,
butterfly bushes are like a
butterfly party!

Some butterflies have bright colors that act like a warning sign telling predators, "Stay away!"

Monarch butterflies fly south
for the winter to find cozy,
warm places to stay!

I hope you had a super fun adventure discovering the secret world of butterflies! Don't forget to spread love, peace, and the magic of nature with everyone!

Fun Fact: Caterpillars munch on leaves, flowers, and seeds as they grow!

Amara's™
Adventures

Amara's™ Adventures